UNDERSTANDING THE THEORIES OF COSMOLOGY AND ASTRONOMY

A SIMPLE EXPLANATION OF VARIOUS THEORIES-FROM THE BIG BANG THEORY TO THE THEORY OF EVERYTHING

KANCHANA MUNIRATHNAM

ISBN 978-1-68538-800-3

Dedicated to our wonderful Universe, scientists who have uncovered its marvels, and my family.

Contents

Contents

UNDERSTANDING THE THEORIES OF COSMOLOGY AND ASTRONOMY

CHAPTER ONE

Introduction

What is the structure of the Universe, How it started, How vast is it, How it will end?

Is the universe now in its infancy, in its prime of life, or its old age? Will it end one day or remain as it is?

It's difficult for most of us to think about these questions- who spend most of the time think about topics like " which TV program do I need to watch today? Or what's for dinner, Politics, Movie stars, our Jobs, bills to pay, etc.

But as a whole- Humankind always pondered on these mind blogging questions and tried to understand the Universe through assumptions, observations, and experiments.

Even the earliest known philosophers and theologians thousands of years ago speculated about the nature and origin of the universe. Historically, they have helped to set the stage for later investigations about initial conditions in the universe.

We humans have discovered that requiring assertions about nature to be testable by observation, experiment, or calculation allows us to gain a better understanding of the universe around us. From ancient times to the modern scientific era, the science of cosmology has progressed and saw many shifts through the application of experiments and mathematical models.

To understand how cosmologists have arrived at their present conceptions of the universe, we will review a series of cosmological and Astronomical theories developed at different times.

This book provides a nonmathematical overview of various theories in cosmology and Astronomy. If a deeper understanding of the subject is desired, more in-depth learning is required.

> "Who are we? We find that we live on an insignificant planet of a humdrum star lost in a galaxy tucked away in some forgotten corner of a universe in which there are far more galaxies than people.
>
> -Carl Sagan"

STEADY STATE THEORY

CHAPTER TWO

STEADY STATE THEORY

The steady-state model asserts that although the universe is expanding, it does not change its appearance over time-the perfect cosmological principle - the universe has no beginning and no end. This required that matter be continually created to keep the universe's density from decreasing. Influential papers on steady-state cosmologies were published by Hermann Bondi, Thomas Gold, and Fred Hoyle in 1948. Similar models had been proposed earlier by William Duncan MacMillan, among others.

At the core of the Steady State, the theory is the Perfect Cosmological Principle.This states that the Universe is infinite in extent, infinitely old and, taken as a whole, it is the same in all directions and at all times in the past and at all times in the future. In other words, the Universe doesn't evolve or change over time.The Steady State theory proposes that new stars are recurrently created all the time at the rate needed to replace the stars which have used up their fuel and have stopped shining. So, if it is taken a large enough region of space, and by large means tens of millions of light-years across, the average amount of light emitted doesn't change over time. Hubble proved that the galaxies are all moving away from each other, which implied that the average distance between galaxies is increasing and so the Universe must be changing over time.

The Steady State theory gets curved this by assuming that new matter is continuously created out of nothing at the incredibly small rate of 1 atom of hydrogen per 6 cubic kilometers of space per year. This new matter eventually forms new stars and new galaxies and, if we take a large enough region of the Universe, its density, which is the amount of matter in a given volume of space, doesn't change over time. If two individual galaxies are taken then their relative distance *will* get further and further apart due to the expansion of the Universe. However, because new galaxies are being formed all the time, the average distance between galaxies doesn't change.

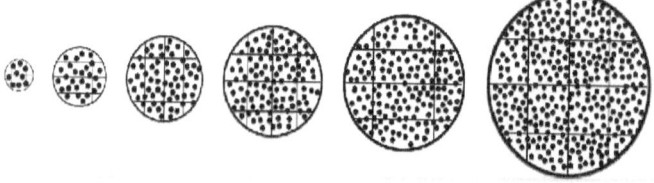

The matter is constantly created as the universe expands

The major goal of the steady-state theory was to explain the expansion of the universe without having to say that the universe as a whole looks different at different points in time. If the universe at any given point in time looks basically the same, there is no need to assume a beginning or an end. Bondi and Gold proposed no mechanism for the creation of matter required by the Steady State Theory, but Hoyle proposed the existence of what he called the "C-field," where "C" stands for "Creation." The C-field has negative pressure, which enables it to drive the steady expansion of the cosmos, whilst also creating new matter, keeping the large-scale matter densityapproximately constant.

Quasi-steady-state cosmology (QSS) was proposed in 1993 by Fred Hoyle, Geoffery Burbidge, and Jayant V.Narlikar as a new incarnation of the steady-state ideas meant to explain additional features unaccounted for in the initial proposal. The model suggests pockets of creation occurring over time within the universe, sometimes referred to as mini bangs, *mini-creation events,* or *little bangs.* After the observation of an acceleration universe, further modifications of the model were made.Both steady-state and quasi-steady-state assert that the creation events of the universe (new hydrogen atoms in the steady-state case) can be observed within the observable universe. The Steady-State model also predicted that the steady creation of antimatter and neutrons would result in regular annihilations and neutron decay, thus leading to the existence of a gamma-ray background and hot, x-ray emitting gas throughout the Universe.

The Steady State theory was very popular in the 1950s. However, the evidence against the theory began to emerge during the early 1960s. Firstly, observations taken with radio telescopes showed that there were more radio sources a long distance away(billions of light-years) from us than would be predicted by the theory.

Another piece of evidence to discredit the theory emerged in 1963 when a new class of astronomical objects called quasars was discovered. These are incredibly bright objects which can be up to 1,000 times the brightness of the Milky Way but are very small when compared to the size of a galaxy. Quasars are only found at great distances from us, meaning that the light from them was emitted billions of light-years ago. The fact that quasars are *only* found in the early Universe provides strong evidence that the Universe has changed over time.

Also, the discovery of extremely active galaxies, in which the accretion of mass onto central, supermassive black holes releases sufficient radiation to outshine the entire galaxy, cemented the evidence against a steady-state universe.Problems with the steady-state model began to emerge when observations began to support the idea that the universe was in fact changing bright radio sources were found only at large distances, not in closer galaxies. Most cosmologists believed that statistical tests based on radio-source surveys had ruled out the steady-state model by 1961, however, some proponents of the steady-state model claimed that the radio data were questionable.

The real setback of the Steady State theory was the discovery in 1965 of cosmic microwave background radiation. This is weak background radiation that fills the whole of space and is the same in all directions. In the Big Bang theory, this radiation is a relic or snapshot from the time the Universe was young and hot and was predicted before it was discovered. However, in the Steady State theory, it is almost impossible to explain the origin of this radiation.

THE BIG BANG THEORY

CHAPTER THREE

THE BIG BANG THEORY

The Big bang theory is the most widely accepted model of the origin of the Universe. The Big Bang theory developed much before from observations of the structure of the universe. In 1912 Vesto Slipher- an American astronomer measured the first Doppler shift of a spiral nebula and discovered almost all such nebulae were receding from Earth, but the cosmological implication of this was not grasped.

This idea of the evolving universe totally deviated from the idea of the Static universe started in 1922 to 1924 when the Russian mathematician Alexander Friedmann developed a set of equations known as the Friedmann equation which showed that the universe may be expanding. In 1927 Belgian cosmologist George Lemaitre published a virtually unnoticed paper that provided the solution for the observed expansion of the universe.

Lemaitre's idea was everything started from 'Primeval Atom' which exploded giving rise to Space and time and expansion of the universe. He explored the consequences of an expanding universe and proposed that the universe must have started at a finite point that disintegrated in an expansion. This idea gave birth to a new expanding universe concept. Lemaitre's picture of a primordial expanding space filled with matter and energy excited cosmologists and physicists as it paved the way for them to study the evolution and observed expansion of the universe. This idea of expanding and evolving universe was developed further by another cosmologist including physicist George Gamow which gave rise to modern Big Bang theory.

Edwin Hubble's observation in 1929 where various Galaxies moved away from us formed the basic proof for expanding the idea of the universe. In his short paper Hubble presented the observational evidence for the expanding universe. His observation showed more that distant galaxies recede faster than nearby galaxies. This crucial observation that the farther away galaxies are, the faster they are moving away from us provided the validity for the Big bang theory. Also, Hubble reasoned that if galaxies are moving away from us, then in past they must have clustered together.

Hubble's law, also known as the Hubble–Lemaître law, is a physical cosmological observation that galaxies move away from the Earth at rates proportionate to their distance. To put it another way, the farther they are from Earth, the quicker they are rushing away from it. The redshift of the galaxies, which is a shift in the light they emit toward the red end of the spectrum, has been used to measure their velocity.

# HUBBLE'S LAW
VELOCITY = HUBBLE CONSTANT x DISTANCE

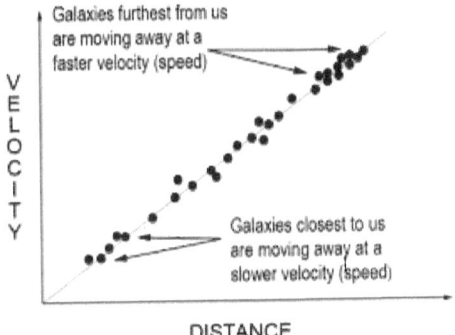

Hubble's law-farther away galaxies are, the faster they are moving away from us

Another important discovery in 1964 by American Radio astronomers Arno Penzias and Robert Wilson which is Cosmic Background Radiation CMB provided another landmark support to Big Bang theory. CMB is electromagnetic radiation which is a remnant from the early stage of the universe. It is uniform thermal energy radiation that fills the entire universe and is considered as a shockwave of the Big bang.

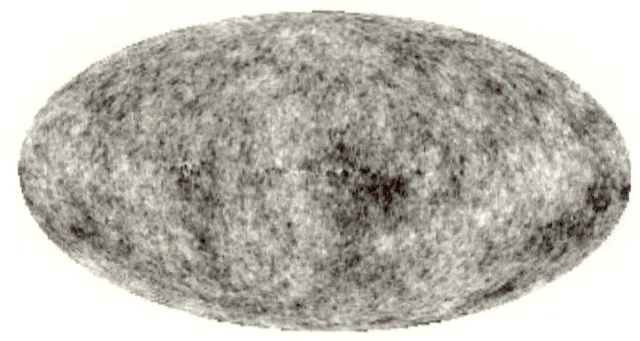

Cosmic background radiation, electromagnetic radiation coming from every direction in the universe

Understanding the Big Bang model

Interestingly, Fred Hoyle was credited with coining the term 'Big Bang' during the talk in 1949 BBC Radio broadcast who in fact favored the steady-state cosmological model. When we think about the Big bang, we always imagine that there was an explosion at one point somewhere in space. But it was a state where everything-Space, Time, and Expansion started simultaneously.

Thus Big bang is not an explosion but an expansion of space which itself expanding with time everywhere. This should not be pictured as matter moving outwards to fill space. The Big bang can be a point that gave birth to our universe as we see now and from where laws of physics can be applied. Observation and measurement of the current state of the universe place the singularity state at around 13.8 billion years which is calculated as the age of the universe. The whole picture of the model is that the universe expanded from an initial state of high density and temperature and cooled sufficiently to allow the formation of subatomic particles and atoms.

Clouds of primordial elements formed early stars and galaxies. After a few million seconds of the Big Bang, after the initial state of extreme temperature quarks aggregated to form protons and neutrons and a few minutes later these protons and neutrons combined to form nuclei. It is predicted that it took millions of years for electrons to get into orbit of the nuclei which gave birth to the first atom.

These atoms were mostly hydrogen and helium which are the most abundant elements in the universe. During the first few minutes of the Big Bang light elements were born in a process known as nucleosynthesis. The universe was essentially too hot for light to shine.

The heat smashed atoms together with enough force to break them up into a dense plasma, an opaque soup of protons, neutrons, and electrons that scattered light like fog. This initial flash of light created is detectable even now as uniform radiation throughout the universe which is Cosmic Background radiation CMB.

About a million years after Big Bang, the universe began to emerge from the cosmic dark age when clumps of gas collapsed to form the first stars and galaxies. In the earliest minutes of the universe, there was no structure to it with matter and energy distributed nearly uniformly throughout.

The gravitational pull of fluctuation in the density of matter gave rise to the vast web-like structure of stars, galaxies, and space has seen today. The expansion of the universe slowed down and the slightly denser regions of the distributed matter gravitationally attracted nearby matter and thus grew into denser, forming gas clouds, stars, and galaxies.

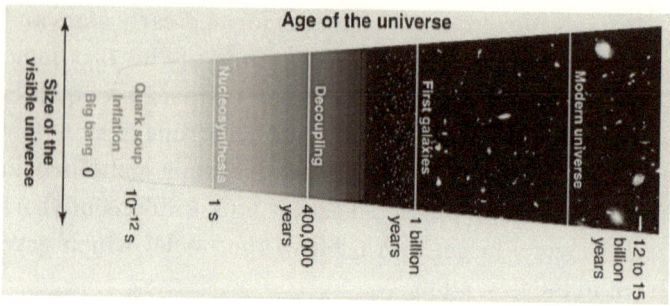

Big Bang Timeline

It is beyond the model of the Big bang to define the encompassing universe that is expanding and what was before Big Bang and what caused it. Although Big Bang theory does not describe how space, energy, and the time came into being, it describes the emergence of the present universe from starting density and temperature and it's beyond our capacity to imagine or replicate the initial state and earliest states of the universe are more speculative. How the initial state of the universe originated? and what was before this state? is still an open question. However, this model explains a broad range of observed phenomena like an abundance of light elements, Cosmic Microwave Background radiation, and large-scale structure.

INFLATION THEORY

CHAPTER FOUR

INFLATION THEORY

Although the Big Bang model explains the CMB, abundance of light elements, there are several characteristics of the universe that cannot be explained using the model.

These characteristics are the temperature of CMB is the same where ever we look which implies this sameness comes from the state that the visible universe was in contact at some point and requires extreme fine-tuning of condition which would be an unbelievable coincidence. When two things have the same temperature, like a coffee in a cup, two substance touching each other, these things has the opportunity to interact and transfer the heat with each other and get their temperature equalized. But imagine that we didn't give enough time for the coffee to be in the cup and poured out of the cup, still the cup has the same temperature as a coffee. So this is the question we have -how can this happen?. Likewise, when we see the opposite regions of the universe, these two regions are having uniform temperatures but didn't have enough time to be in contact and exchange temperature.

Assuming the standard Big Bang model, all distant regions in the universe could never have been in causal contact with each other because the light travel time between them exceeds the age of the universe. And also to explain the observed homogeneity of causally disconnected regions of vast space in the absence of a mechanism that sets the same initial conditions everywhere. This problem is called The Horizon problem.

Another characteristic of the universe is Critical density which was necessary for the universe we see observe now. Initial density would have been high or low, that the expansion would have been too rapid or slow to form any stars, galaxies, and peculiar geometric flatness at large scales. Flat in the sense that parallel lines will remain parallel forever as they travel through the universe. Under the Big Bang model, it is expected the curvature of the universe grows with time. This is called the Flatness problem

One more observation that could not be explained with standard Big Bang theory is the Monopole-single, lonesome north or south magnetic pole, that with just one magnetic pole instead of two, predicted to exist in the early universe. Stable magnetic monopoles should have been formed in the early universe, but these monopoles are never observed as predicted by Big Bang Theory. Modern particle theory predicts that large quantities of superheavy particles that carry magnetic charge should have been created in the initial condition of extreme temperature and density during the explosive birth of the universe as by the Big Bang Model.

These particles should be very stable and should be still around, however, there is no evidence of such particles and not been observed till now. This is called the Monopole problem which could not be explained by the Big Bang model. The Inflation Theory which was developed by Alan Guth, Andrei Linde, Paul Steinhardt, and Andy Albrecht in the 1980s offered solutions to these problems.

The model describes a period of extremely rapid expansion of the universe before the gradual Big Bang expansion. Inflation was both rapid and exponential which increased the size of the universe more than a factor of 10^26 in a fraction of a second. This event known as Cosmic inflation explains many features found in astronomical observation.

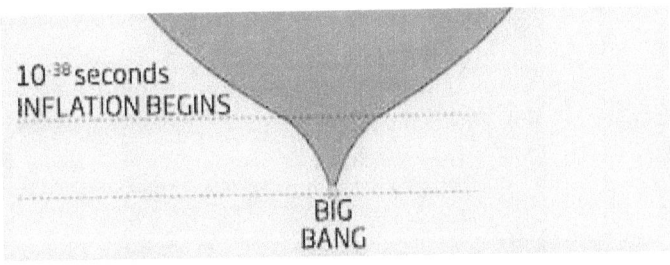

Inflation

The model proposes that the initial exponential expansion of the universe which magnified the quantum fluctuations that were created at the beginning led the way to form galaxies. This quantum variation which was amplified during Inflation acted as seeds for structure formation as the density of matter got varied. Gravity caused more dense regions to start contracting to lead to the formation of stars and galaxies. In other words, Quantum fluctuations in the microscopic inflationary region magnified to cosmic scale led to the start of structure formation in the universe.

An analogy could be if small microscopic dots are drawn on a balloon and when the balloon is inflated, the dots would grow bigger and bigger which means that inflation acts as a microscope that magnifies the dots on the balloon. In the same way expansion of the universe during the inflationary epoch magnified the quantum fluctuation which left imprints in the CMB and in the distribution and formation of galaxies as observed now.

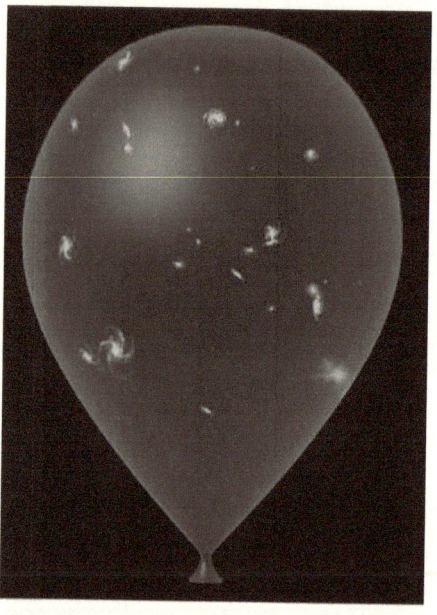

Expansion of universe-inflationary epoch

The Universe's expansion during the inflationary period acts as a massive microscope, magnifying quantum fluctuations. The cosmic microwave background radiation (hotter and colder regions) and the dispersion of galaxies bear witness to this.

How the Theory answered the problems in the Big Bang model:

Flatness problem:The exponential expansion during the inflation epoch triggered enough expansion to make the universe appear as flat answers observed geometric flatness at a large scale of the universe. Even if the universe started had curvature when it started, this huge inflation would cause the universe to appear flat as we observe today. An analogy could be that the earth is large enough that it appears to be flat to us even though the surface we stand on is curved outside of the sphere. The Inflation theory predicts that the ultrafast inflation would have expanded away any large curvature of the universe. Thus this model solves the problem of the observed flatness of our universe.

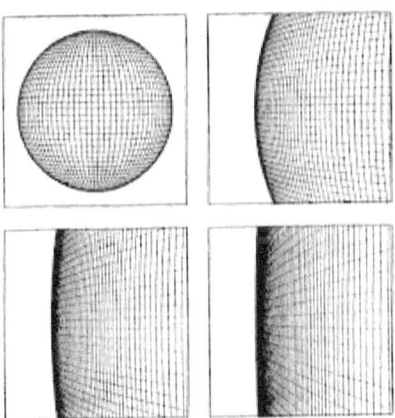

The inflation stretched the initial curvature of the 3-dimensional universe to near flatness.

Horizon Problem: As Inflation gives an exponential expansion in the early universe before the actual Big Bang started, it is that distant regions in the universe we observe today were actually much closer together and thus giving the uniformity we observe now. This theory predicts that regions that appear to be isolated from each other were in contact before the inflation period and could have attained the uniform temperature we observe in Cosmic Micro Background radiation CMB.

As per the inflationary model, the universe increased in size by a factor of more than 10^26, from a small and causally connected region in near equilibrium. Inflation then expanded the universe rapidly, isolating nearby regions of space-time by growing beyond the limits of casual contact, effectively giving the uniformity at a large distance. The model suggests that the universe was entirely in contact with the very early universe.

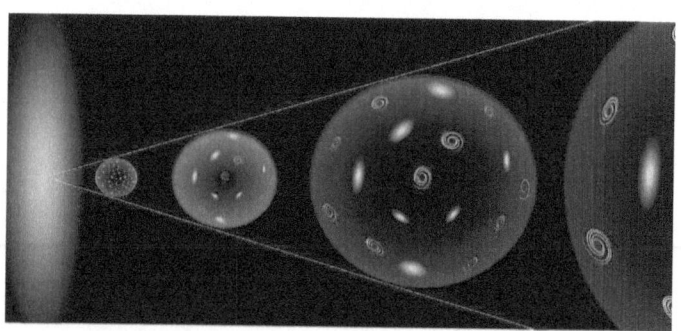

Regions in Universe were close together

We observe the CMB after inflation has occurred at a very large scale. So it maintained thermal equilibrium to this large size because of the rapid expansion from inflation and difference in the temperature of CMB are smoothed by inflation, thus solving the Horizon problem in the standard Big Bang model.

Monopole problem:The Inflation theory solves the other problem which is the formation of magnetic monopoles, where it says initial expansion dilutes exotic heavy particles like magnetic monopoles. The inflation model allows magnetic monopoles to exist as long as they were produced before the period of inflation. During inflation, the density of monopoles drops exponentially, so their abundance drops to undetectable levels.

The model proposes that as a result exponential expansion during the inflation epoch puts so much space between the individual monopoles that detection becomes highly impossible. Thus monopoles were produced, then during inflation, they have driven away from each other thereafter diluting their density to un observably low levels.

When inflation ended and the universe got hot again, it never achieved the ultra-high temperature necessary to create them again, thus answering the absence of monopole in the universe. Even though the Inflation model explains and offers a solution to observed homogeneity and geometric flatness of the universe, there is no explanation for the critical initial condition. The model cannot be evaluated scientifically. Also expected outcome of inflation can easily change if we vary the initial conditions. Inflation requires an extremely specific initial condition of its own so that the problem of the initial condition is not solved.

However, a team of astronomers led by John Kovac has detected a polarization pattern in the microwave background radiation that is the first direct evidence for cosmic inflation. This B- mode polarization is the pattern created by gravitational waves. The 10-meter south pole telescope BICEP(Background Imaging of Cosmic Extragalactic Polarization) telescope against the Milky way recently detected gravitational waves in the cosmic microwave background, a discovery that supports the cosmic inflation theory.

The data also represent the first images of gravitational waves or ripples in space-time. These waves are described as the first tremors of the Big Bang and are a smoking gun for inflation as alternative theories do not predict such a signal.BICEP2's measurement of inflationary gravitational waves is an impressive combination of theoretical reasoning and technology. Inflation theory, thus a wonderfully productive model that solved some real problems in the observed Universe, and as evidence is discovered, the theory gains confidence among cosmologists.

MULTIVERSE THEORY

CHAPTER FIVE

MULTIVERSE THEORY

Most of the hundreds of models spawned by Dr. Guth's original vision suggest that inflation, once started is eternal. Even as our universe settled down to a comfortable expansion with atoms, stars, and galaxies, the rest of the cosmos will continue blowing up, spinning other bubbles here and there endlessly, and paving a way to a concept known as the Multiverse.

The Multiverse is a controversial idea that describes that our Universe and all that is contained within is just one small part of a larger structure. The larger structure encapsulates our observable universe as a small part of a larger universe that extends beyond the limits of our observations. This unobservable universe may include many universes that are disconnected from each other and which may or may not be similar to the universe we live in.

The multiverse is a fictitious collection of multiple universes. These universes combine to form everything that exists: space-time, matter, energy, and the physical laws and constants that describe them. The various universes within the multiverse are referred to as "parallel universes" or "alternate universes."

Bubble universe-Larger structure encapsulates our observable universe

Because the observable universe extends only as far as the light had a chance to get in the 13.8 billion years since the Big Bang, the spacetime beyond that distance can be considered to be its own separate universe. In this way, a multitude of universes exists next to each other. In addition to the multiple universes created by extending space-time, other universes may arise from a theory called 'eternal inflation. Eternal inflation is first proposed by cosmologist Alexander Vilenkin, who suggests that some pockets of space stop inflating, while other regions continue to inflate, thus giving rise to many isolated Bubble universes.

Thus our universe where inflation has ended allowing to form of stars and galaxies is just a small bubble in a larger structure in which some regions are still inflating and the physics and fundamental constants may differ from our universe. The universes that may exist can be considered universes that may be similar to our universe or may have the same fundamental laws of physics but started with different initial conditions. For Multiverse to exist we need inflating Universe and quantum mechanics as support. Inflation is treated as a field, like all the quanta in the universe obeying the rules of quantum field theory. The rule governing quantum uncertainty is the one needed to support the existence of the Multiverse.

When the universe inflates, the value of the field changes slowly. In different inflating regions, the field value spreads out randomly with different amounts and in different directions. In some regions, inflation ends quickly, in others, it ends more slowly. This is the important point that describes the existence of the Multiverse. When inflation ends, we get a hot Big bang and a large universe, whereas a small part of it might be similar to our own observable universe. There may be some other regions where inflation continues for longer and giving rise to another large hot Big Bang and large Universe. Some other regions are not just inflating but also growing. So new universes are formed by many inflating regions.

Thus wherever inflation occurs, it gives rise to exponentially more regions of space with each step forward in time. Even if there are regions where inflation ends, there are far more regions where inflation will continue. This process never comes to an end is what makes inflation eternal once it begins and this gives rise to the notion of a Multiverse. These huge universes, bigger than the small part that is observable to us constantly being created is the concept of Multiverse and a theoretical consequence that comes from Inflation and quantum mechanics. The inflationary universe governed by quantum mechanics makes Multiverse unavoidable.

Max Tegmark's four levels:

Max Tegmark, a cosmologist, has created a taxonomy of universes beyond the known observable universe. Tegmark's classification is divided into four levels, each of which can be considered to contain and develop on the preceding levels. Below is a brief description of them.

Level I: An extension of our universe

The presence of an infinite ergodic universe, which, since it is infinite, must contain Hubble volumes(a spherical region of the observable universe surrounding an observer beyond which objects recede from that observer at a rate greater than the speed of light due to the expansion of the Universe) realizing all beginning conditions, is a prediction of cosmic inflation.

As a result, an infinite universe will have an endless number of Hubble volumes, all of which will have identical physical laws and constants. Almost all configurations, such as matter distribution, will be different from our Hubble volume. However, because there are an endless number of Hubble volumes well beyond the cosmological horizon, there will be Hubble volumes with comparable, if not identical, configurations at some point.

Level II: Universes with different physical constants

The multiverse or space as a whole is stretching and will continue to do so indefinitely, according to the eternal inflation theory, yet some parts of space cease stretching and form separate bubbles like gas pockets in a foam. These bubbles are multiverses at the embryonic level I. Different bubbles may suffer distinct levels of spontaneous symmetry breakdown, resulting in different physical constants and behaviors. Level II also includes the oscillatory universe theory of John Archibald Wheeler and the theory of the fecund world of Lee Smolin.

Level III: Many-worlds interpretation of quantum mechanics

In a nutshell, one of quantum mechanics' tenets is that certain observations cannot be anticipated with absolute certainty. Instead, there are several possibilities, each with a different probability. Each of these possible observations, according to the MWI, belongs to a separate universe. Assume that a six-sided die is thrown and that the outcome is observable in quantum physics. Six separate worlds correspond to each of the six potential outcomes of the dice.

If there were an infinite number of universes, each with its own set of physical laws or fundamental physical constants, some of them even if only a few would have the right combination of laws and fundamental parameters for the development of matter, astronomical structures, elemental diversity, stars, and planets that can exist for long periods of time.

Recent developments in particle, quantum mechanics, and cosmology lead naturally to the postulate of a multiverse. Although direct confirmation of other universes or regions of our universe, maybe infeasible or even impossible in principle, the multiverse theory does make some observable predictions and can be tested in the future.

THEORY OF RELATIVITY

CHAPTER SIX

THEORY OF RELATIVITY

The theory of relativity incorporates two interrelated theories by Albert Einstein special relativity and General relativity. General relativity enlightens the law of gravitation and its relation to other forces of nature and Special relativity applies to all physical singularities in the absence of gravity. The theory largely altered theoretical physics and astronomyduring the 20th century, prevailing a 200-year-old theory of mechanics created primarily by Isaac Newton.

Special Theory of Relativity

The theory of special relativity was developed by Albert Einstein in 1905, and it forms part of the basis of modern physics. The theory of special relativity explains how space and time are linked for objects that are moving at a consistent speed in a straight line. Einstein's special theory of relativity(special relativity) is all about what's relative and what's absolute about time, space and motion.

One of the most surprising features of special relativity is that several statements and results which usually thought to be absolute became observer-dependent. In particular, statements about space and time, distances, and duration turn out to be relative.

One of the most famous equations in mathematics comes from special relativity. The equation — $E = mc^2$ — means "energy equals mass times the speed of light squared." It shows that energy (E) and mass (m) are interchangeable; they are different forms of the same thing. If the mass is somehow totally converted into energy, it also shows how much energy would reside inside that mass. This equation is one of the demonstrations of why an atomic bomb is so powerful, once its mass is converted to an explosion.

This equation also shows that mass increases with speed, which effectively puts a speed limit on how fast things can move in the universe. The speed of light (c) is the fastest velocity at which an object can travel in a vacuum. As an object moves, its mass also increases. Near the speed of light, the mass is so high that it reaches infinity, and would require infinite energy to move it, thus capping how fast an object can move. The only reason light moves at the speed it does is because photons, the quantum particles that makeup light, have a mass of zero.

For example, simultaneity is a relative concept. Imagine that there are two events that an observer in space station A judges to be simultaneous – say, the explosion of an object at one point in space and spilling of coffee going off a few miles away. For an observer in space station B, which is moving relative to A, this statement will not necessarily be true: In general, such an observer will conclude that one of the events happened earlier than the other. Similarly, the temporal duration depends on the observer. This relativistic effect is called *time dilation*. Summarized moving clocks are slower than stationary ones. An observer on station A measures time using his onboard clock. Station B, passing A at high speed, has an exact copy of A's clock on board. Yet, from the point of view of A, the clock in station B runs more slowly than his own.

In special relativity, simultaneity is relative, and so are time and space. Observers moving relative to one another come to different conclusions about which events happen simultaneously. They agree only about what events there are, not about where or when these events take place. *Spacetime,* the totality of all events, is absolute. But there are different ways to slice that collection of events into snapshots of simultaneity. When viewed in succession, these snapshots show how, with time, changes take place in space. Different observers, looking at different successions of snapshots, come to different conclusions about which events are simultaneous. Space-time is absolute, space and time are not.

Nevertheless, space-time does retain what physicists call *a causal structure,* which is the same for all (inertial) observers. This structure determines which events can, in principle, be influenced by which other events, and also for which pairs of events mutual influence is impossible. By concluding, moving clocks tick at a slower rate, light speed is the same for all (inertial) observers, and lengths and distances depend on who measures them.

Thus Special relativity is a theory in physics that concerns the relationship between space and time and says that they're two sides of the same coin: space-time. Like all scientific theories, it is backed by a large body of evidence and is widely accepted as being accurate.

The two main postulates of special relativity are:

- The laws of physics are the same in all reference frames that are moving at a constant velocity (not accelerating).
- The speed of light is the same in all of these reference frames, even if the source of the light is moving.

Thus Special relativity is a theory of the structure of spacetime.

General Theory of Relativity

General relativity, often known as the general theory of gravitation, is Albert Einstein's geometric theory of gravity, published in 1915.

General relativity differs from those of classical physics, especially concerning the passage of time, the geometryof space, the motion of bodies in free fall, and the propagation of light. Examples of such differences include gravitational time dilation, gravitational lensing,the gravitational redshift of light, the gravitational time delay,and singularities /blackholes.

General Relativity can be entitled as Theory of Gravity. Gravity-instead of being an invisible force that attracts objects to one another is a curving or warping of space. The more massive an object, the more it warps the space around it. For example, the sun is massive enough to warp space across our solar system same as the way a heavy ball resting on a rubber sheetwarps the sheet.

As a result, Earth and the other planets move in curved paths (orbits) around it. Just as gravity can stretch or warp space, it can also dilate time.Thus Gravity is a distortion of space (or more precisely, spacetime) caused by the presence of matter or energy. A massive object generates a gravitational field by warping the geometry of the surrounding spacetime.

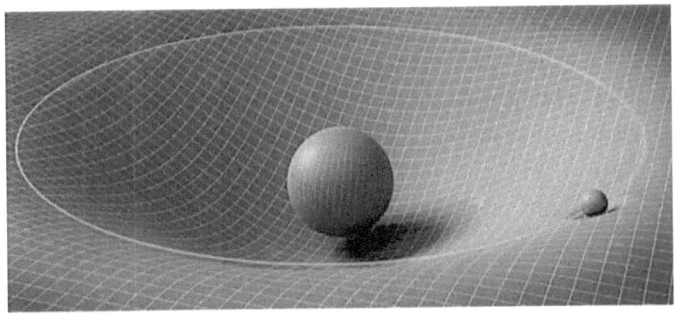

Space-time curvature-massive object generates a gravitational field by warping the geometry of the surrounding spacetime.

Space was an abstract concept, a kind of container, not a perceptible entity that could effect change. Einstein realized that time could warp. Intuitively, it is always envisioned that clocks, regardless of where they are located, tick at the same rate. But Einstein proposed that the nearer clocks are to a massive body, like the Earth, the slower they will tick, reflecting a startling influence of gravity on the very passage of time. And much as a spatial warp can bump an object's trajectory, so too for a progressive one. Einstein's math suggested that objects are drawn toward locations where time elapses more slowly.

According to general relativity, light does not travel along straight lines when it propagates in a gravitational field. Instead, it is deflected in the presence of massive bodies. In particular, starlight is deflected as it passes near the Sun, leading to apparent shifts of up 1.75 arc seconds in the stars' positions in the sky.

General relativity has been established as an indispensable tool in modern astrophysics. It provides the foundation for the current understanding of black holes, regions of space where the gravitational effect is strong enough that even light cannot escape. Their strong gravity is thought to be responsible for the intense radiation emitted by certain types of astronomical objects such as active galactic nuclei or microquasars. General relativity is also part of the framework of the standard Big bang model of cosmology.

One of the most important aspects of general relativity is that it can be applied to the universe as a whole. A key point is that, on large scales, our universe appears to be constructed along very simple lines: all current observations suggest that, on average, the structure of the cosmos should be approximately the same, regardless of an observer's location or direction of observation. Such comparatively simple universes can be described by simple solutions of Einstein's equations.

In contrast to all other modern theories of fundamental interactions, general relativity is a classical theory: it does not include the effects of quantum physics. The mission for a quantum version of general relativity addresses one of the most fundamental open questions in physics.

General relativity has emerged as a highly successful model of gravitation and cosmology, which has so far passed many unambiguous observational and experimental tests. However, there are strong signs the theory is incomplete. The problem of quantum gravity and the question of the reality of spacetime singularities remain open. These singularities are boundaries ("sharp edges") of spacetime at which geometry becomes ill-defined, with the consequence that general relativity itself loses its predictive power.

Observational data used to support the existence of dark energy and dark matter could point to the need for new physics. Even if taken at face value, general relativity offers a plethora of avenues for additional investigation. General relativity is still a hot topic of research a century after it was first proposed. Broadly accredited as a theory of astonishing beauty, general relativity has often been described as the most beautiful of all existing physical theories.

The theory of relativity, according to Einstein, belongs to a category of "principle-theories." As a result, it adopts an analytic technique, which means that the pieces of this theory are founded on empirical discovery rather than supposition. We comprehend the broad characteristics of natural processes by seeing them, developing mathematical models to describe what we saw, and deducing the essential conditions through analytical tools. Separate event measurements must meet these criteria and agree with the theory's findings.

STRING THEORY

CHAPTER SEVEN

STRING THEORY

String Theory is a theoretical framework that endeavors to address many questions and fundamental problems in the physics of atomic nuclei, black holes, and the early universe. The framework describes that point-like particles of particle physics are replaced by one-dimensional objects called strings. It explains how these strings propagate through space and interact with each other.

A string looks just like an ordinary particle, with its mass, charge, and other properties determined by the vibrational state of the string. The core of the idea is that fundamental particles observed are not point-like dots, but are tiny strings. According to string theory, the matter is made up of strings. What is perceived as particles are actually vibrations in the loops of string, each with its own characteristic frequency.

String Theory says that the fundamental particles are extremely small one-dimensional objects known as strings. Strings are the smallest thing in the whole universe as everything, even the subatomic particles are made up of these strings. According to the theory, all fundamental particles are actually tiny vibration loops of string. Various properties of particles like mass and charge are determined by the vibrations of the one-dimensional string-like entities. The size of the string is predicted to be about 10^-33cms. With an analogy, if atoms were magnified, to the size of the whole observable Universe, then the String would be the size of a small house.

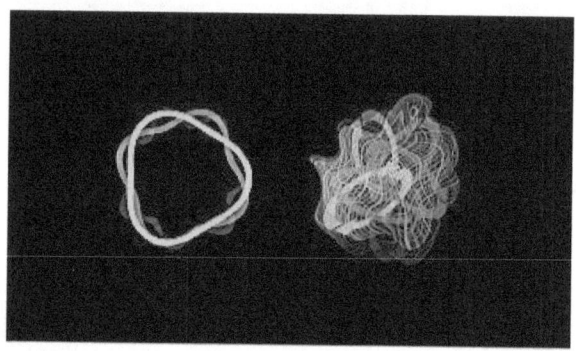

Tiny vibrating strings

Tiny vibrating strings that twist and turn in complicated ways look like particles. A string of a particular length striking a particular length note gains the properties of a photon, and another string folded and vibrating with different frequency plays the role of a quark, and so on. The Framework proved striking for its potential to explain fundamental constants like the mass of the electron. All the fundamental forces except gravity are described within the framework of quantum mechanics. According to Einstein's general theory of relativity, gravity is the distortion of space-time caused by the presence of matter. So a quantum theory of gravity was needed to reconcile General relativity with the principle of quantum mechanics.

String theory helps here and provides a unified description of gravity and particle physics. One of the many vibrational states of the string corresponds to the graviton, a quantum mechanical particle that carries gravitational force. Thus string theory is a theory of quantum gravity.

There are several versions of superstring theory Type I, Type IIA, Type IIB and two flavors of Heterotic string theory. The different theories allow different types of strings, and the particles that arise at low energies exhibit different symmetries.

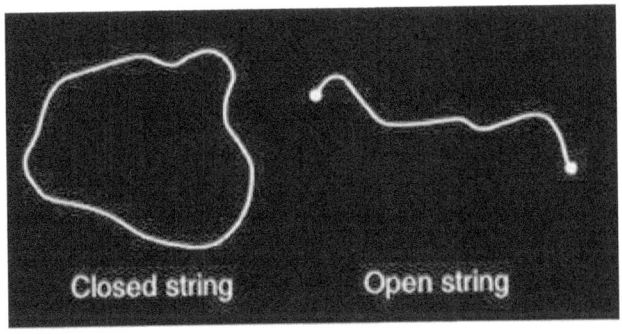

Closed string   Open string

The distinctions between these theories are mathematically sophisticated.

Type I string theory:

Type I string theory involves both open and closed strings. It contains a form of symmetry that's mathematically designated as a symmetry group.

Type IIA string theory:

Type IIA string theory involves closed strings where the vibrational patterns are asymmetrical, regardless of whether they travel left or right along the closed string.

Type II open strings are attached to structures called D-branes with an odd number of dimensions.

Type IIB string theory:

Type IIB string theory involves closed strings where the vibrational patterns are asymmetrical, depending upon whether they travel left or right along the closed string. Type IIB open strings are attached to D-branes with an even number of dimensions.

Brane is a physical object that generalizes the notion of a point particle to higher dimensions. A point particle can be viewed as a brane of dimension zero, while a string can be viewed as a brane of dimension one. It is also possible to consider higher-dimensional branes. Branes are dynamical objects which can propagate through space-time according to the rules of quantum mechanics and have mass and have a charge. The letter "D" in D-brane refers to a certain mathematical condition on the system known as the Dirichlet boundary condition. In string theory, D-branes are an important class of branes that arise when one considers open strings.

Heterotic String theory :

In Heterotic string, the string vibrations in different directions resulted in different behaviors. Left-moving vibration resembled the old bosonic string, while right-moving vibrations resembled the Type II strings.

Two types of this Heterotic theory are Type HO string theory and Type HE string theory.

The earlier version of Bosonic string theory incorporates only the class of particles known as bosons, and which was later developed into superstring theory, which suggests a connection called supersymmetry between bosons and the class of particles called fermions. These five consistent versions of string theory form the limiting case of a single theory in 11 dimensions known as M-theory.

The main feature of String theories is that they require extra dimensions of space-time for their mathematical consistency. It requires extra dimensions of space and contains ways of relating large extra dimensions to small ones. In Bosonic string theory, spacetime is 26-dimensional, and in Superstring theory it is 10-dimensional, and in M- Theory it is 11-dimensional.To describe real physical phenomena using string theory, one must therefore imagine scenarios in which these extra dimensions would not be observed in experiments.

Compactification is one way of modifying the number of dimensions where some of the extra dimensions are assumed to 'close upon themselves to form circles. Another scenario is called brane-world where it is assumed that the observable universe is a four-dimensional subspace of higher-dimensional space.

Compactification is an extension of Kaluza–Klein theory in string theory. It aims to bridge the gap between the four observable dimensions of our universe and the ten, eleven, or twenty-six dimensions that theoretical equations lead us to believe the universe is made up of.

Compactification

String theory has been successful in explaining many complex phenomena, most importantly black holes. As it has predicted the quantization of gravity with gravitons, It becomes a candidate for a unified approach and theory of everything. String theory aims to address various theoretical mysteries, the most fundamental of which is how gravity works for tiny objects like electrons and photons.

One of the difficulties with string theory is that it does not have an acceptable definition in all cases. Another concern is that string theory is expected to explain a vast landscape of possible universes, complicating efforts to construct particle physics theories based on it. Some in the community have criticized these physics approaches and questioned the value of continuing research on string theory unification as a result of these concerns.

String theory exists now for more than thirty years and did not make the least empirically testable prediction. Although String theory gives an explanation for the existence and the number of the particle generation, is certainly not a physical theory, and the rich structure it proposes is not yet understood. The only possible justification of existence lies in the unification program of physics. The existence of gravitons, as well as bosons, is only a promising feature if string theory plays the role of an all-encompassing unified theory.

To test the string theory, physicists need to filter through the massive possible number of solutions to find a manageable amount that may describe our universe.

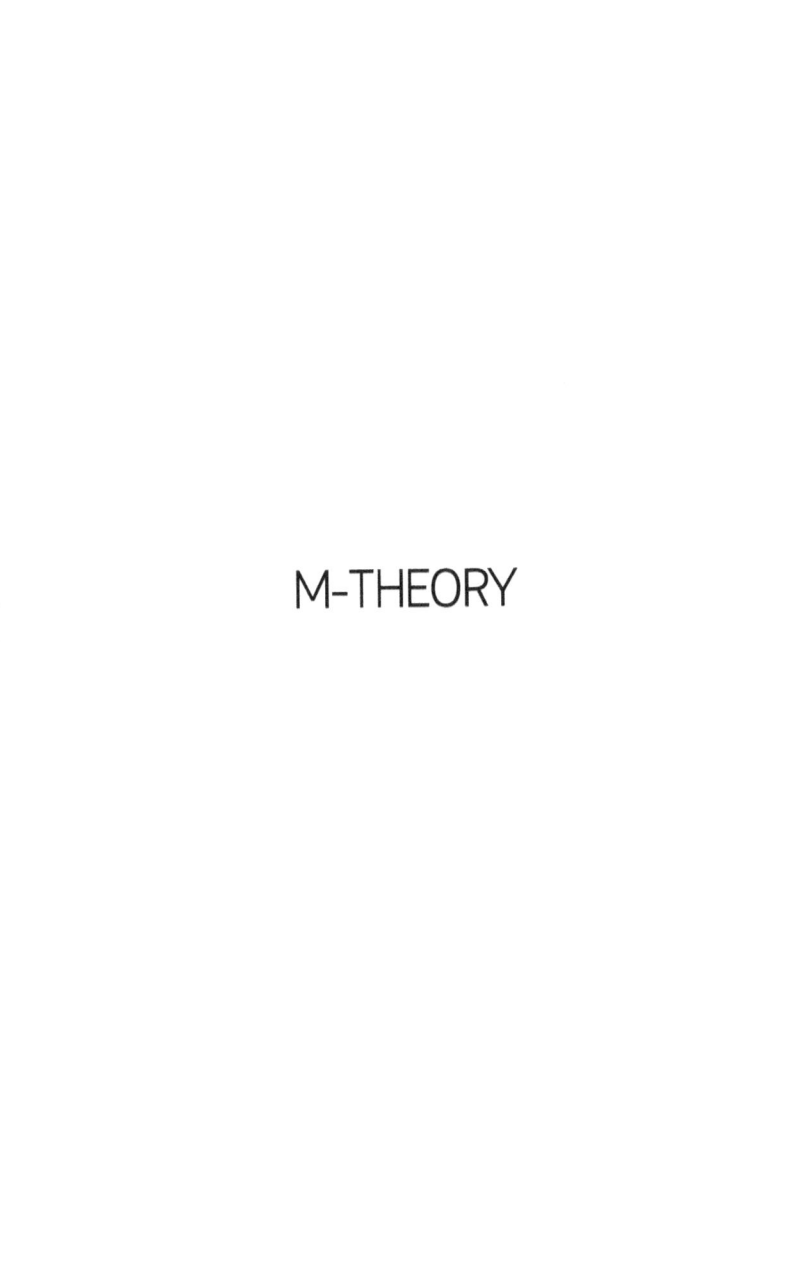

M-THEORY

CHAPTER EIGHT

M-THEORY

M-theory unifies in a single mathematical structure all five consistent versions of string theory. In 1995, the physicist Edward Witten discovered the mother of all string theories which is M-theory.

The five different versions of string theories had few commonalities, they all involved strings. The five versions of string theory required the universe to have total dimensions, the usual 3 spatial dimensions, one for time, and six more compact dimensions that are tiny and curled up on themselves at microscopic scales. In all the String theories the ways strings vibrate give rise to the richness of our physical world, but when it comes to physical theories, details matter and the five competing string models differed in the details. Some only had closed loops of strings and others allowed open strings, some others were allowed to travel in one direction, while others allowed both directions.

As all cannot be correct descriptions of nature, only approximations are available and there is no way of being able to decide which one is best. All these theories appeared to be workable and five contradictory sets of equations cannot describe the same thing. It was assumed that five different versions of string theory might be describing the same thing seen from different perspectives. The unified theory called M- the theory was proposed where 'M' is not specifically defined but is generally understood to stand for-'Membrane',' Matrix',' Master',' Mother'.M-Theory brought all of the string theory together by asserting that strings are really one-dimensional slices of two-dimensional membrane vibrating in 11-dimensional space-time. The combination is accomplished by knitting together a web of relationships between each of the string theories call dualities specifically called S-duality, T-duality, and U-duality.

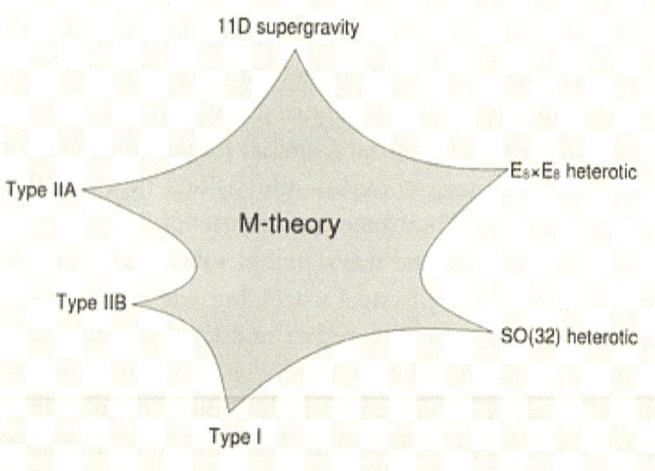

M-theory unifies in a single mathematical structure all five consistent versions of string theory

Each of these dualities provides a way of converting one of the string theories into another. It was seen that by taking a TypeIIA string theory that has a size R and changing the radius to1/R the result will end up being what is equivalent to Type IIB theory of size R. This duality, along with the others, creates connections between all 5 theories.

M-theory shows how all five string theories are really just small corners of a much larger and much more mysterious membrane, where the fundamental object of reality is no longer the strings but the D-brane. Brane is just a multidimensional vibrating thing, with 'D' signifying the dimensions, giving everything from 1-brane(strings) to 2-branes (sheets) to 3-branes(blobs) and more and these branes lie low and mostly just act like strings with 11[th]dimension not playing much of a role. The feature of M-theory predicts the existence of the graviton, a spin-2 particle that mediates the gravitational force. M-theory aims to unify quantum mechanics with general relativity's gravitational force in a mathematically consistent way. One approach to formulating M-theory and its properties is provided by the anti-de sitter/conformal field theory(AdS/CFT) correspondence, proposed by Juan Malsacena. Anti-de Sitter space is a mathematical model of space-time in which the notion of distance between two points is different from the notion of distance in ordinary geometry.

It is closely related to hyperbolic space, which can be viewed as a disk. Three–dimensional anti-de sitter space is like a stack of hyperbolic disks, each one representing the state of the universe at a given time. One can study theories of quantum gravity such as M-theory in the resulting space-time. Another realization of the AdS/CFT correspondence states that M-theory is equivalent to a quantum field theory called ABJM theory which was introduced by Aharony, Bergman, Jafferis, and Maldacena which states that seven of the dimensions of M-theory are curled up, leaving four non-compact dimensions.

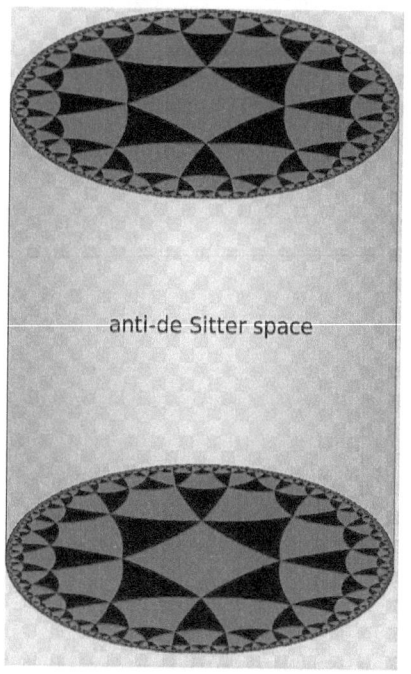

anti-de Sitter space

Anti-de Sitter space

As the space-time of the universe is 4 dimensional, this version of the correspondence provides a somewhat more realistic description of gravity.

One more approach in M-theory pioneered by Witten, Horava, BurtOvrut, and others is called Heterotic M-Theory where one imagines that one of the eleven dimensions of M-theory is shaped like a circle. If this circle is very small, then the space-time becomes effectively ten-dimensional. This model has been used to construct models of brane cosmology in which the observable universe is thought to exist on a brane in a higher-dimensional space.

M theory is not complete, but the mathematical approach has been explored in great detail. So far no experimental support is available for M-theory. Speculatively, M-theory may provide a framework for developing a unified theory of all the fundamental forces of nature. M-theory's mathematical structure has yielded significant theoretical breakthroughs in physics and mathematics. M-theory may, more speculatively, provide a foundation for constructing a unified theory of all of nature's fundamental forces. Attempts to link M-theory to experiment typically focus on compactifying its extra dimensions to build candidate models of the four-dimensional universe, albeit none have been validated to produce physics as seen in high-energy physics experiments.

BLACK HOLES

CHAPTER NINE

BLACK HOLES

Objects with gravitational fields that are too powerful for light to escape were initially considered by John Michell and Pierre-Simon Laplace in the 18th century.Karl Schwarzschild discovered the first modern solution of general relativity that would characterize a black hole in 1916, and David Finkelstein published the first interpretation of it as a region of space from which nothing can escape in 1958.

The term "black star" was coined by John Michell, and the term "gravitationally collapsed object" was used by physicists in the early twentieth century. Marcia Bartusiak, a science writer, attributes the term "black hole" to physicist Robert H. Dicke, who apparently related the event to the Black Hole of Calcutta, a renowned prison where individuals entered but never left alive in the early 1960s.

A student is said to have coined the term "black hole" during a presentation by John Wheeler in December 1967, and it has since become the accepted word for a region of spacetime where gravity is so intense that nothing—no particles or electromagnetic radiation such as light—can escape.

When big stars collapse after their life cycle, black holes of stellar mass develop. A black hole can continue to develop after it has formed by absorbing mass from its surroundings. Supermassive black holes with millions of solar masses may arise by absorbing other stars and merging with other black holes. The existence of supermassive black holes in the centers of most galaxies is widely accepted.

The development of stellar-mass black holes is thought to be caused by the gravitational collapse of massive stars. The development of stars in the early cosmos may have resulted in extremely massive stars that, upon collapsing, formed black holes. These black holes could be the progenitors of the supermassive black holes seen in most galaxies' centers.

When an object's internal pressure is inadequate to resist its own gravity, gravitational collapse happens. This happens when a star has insufficient "fuel" to sustain its temperature through stellar nucleosynthesis, or when a star that would otherwise be stable obtains more matter in a way that does not raise its core temperature. In either situation, the temperature of the star is no longer sufficient to keep it from collapsing under its own weight.A black hole can continue to develop by absorbing more matter once it has been created. Any black hole will collect gas and interstellar dust from its surroundings constantly.

The interaction of a black hole with other stuff and electromagnetic waves such as visible light can be used to infer its presence. Quasars are some of the brightest objects in the cosmos, formed when matter falls upon a black hole and forms an exterior accretion disc heated by friction. Stars that pass too close to a supermassive black hole can be shredded into dazzling streamers before being swallowed. The orbits of other stars around a black hole can be utilized to determine the black hole's mass and location if there are any.

All black hole solutions of the Einstein-Maxwell equations of gravitation and electromagnetism in general relativity may be entirely described by only three externally observable classical parameters: mass, electric charge, and angular momentum, according to the no-hair theorem.Once a black hole has reached a stable state following formation, it has only three independent physical properties: mass, electric charge, and angular momentum; otherwise, the black hole is featureless.

These characteristics are unique in that they can be seen from outside a black hole. A charged black hole, for example, repels other charged objects just like any other charged object. Similarly, the total mass inside a sphere containing a black hole can be calculated far away from the black hole using the gravitational analog of Gauss's law.

When an object enters a black hole, any information about the object's shape or charge distribution is evenly dispersed throughout the black hole's horizon and lost to outside observers. There is no way to avoid losing information about the beginning conditions because a black hole eventually achieves a stable state with only three parameters: the gravitational and electric forces of a black hole reveal very little information about what went in. Static black holes are black holes that contain mass but no electric charge or angular momentum. After Karl Schwarzschild, who discovered this solution in 1916, these black holes are often referred to as Schwarzschild black holes.

Properties and Structure of a Blackhole:

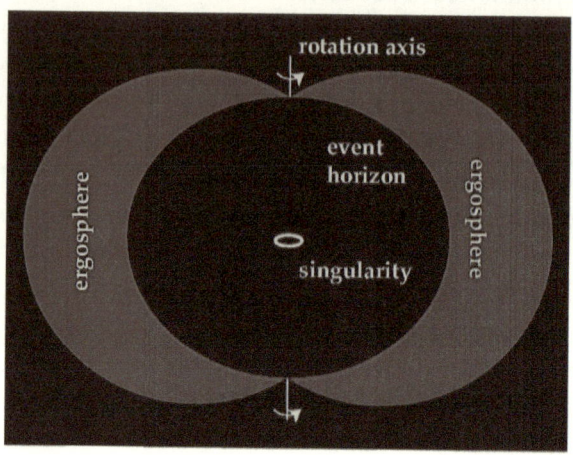

Structure of a Blackhole

Event horizon:

The appearance of an event horizon—a border in space-time through which matter and light may only pass inward towards the black hole's mass—is a defining feature of a black hole. Inside the event horizon, nothing, not even light, can escape. The event horizon is so named because the information from an event that occurs within the border cannot reach an outside observer, making it difficult to tell whether such an event occurred. The existence of a mass deforms spacetime in such a way that particle pathways bend towards the mass, as predicted by general relativity. This deformation becomes so extreme at a black hole's event horizon that no pathways away from the black hole exist.

Clocks near a black hole appear to tick more slowly than clocks further away from the black hole to a distant observer. An object falling into a black hole appears to slow down as it approaches the event horizon, taking an endless amount of time to reach there due to this effect, known as gravitational time dilation.

From the perspective of a fixed outside observer, all processes on this object slow down at the same time, causing any light emitted by the object to look redder and dimmer, a phenomenon known as gravitational redshift. The falling object eventually fades away until it is no longer visible. This process typically occurs very quickly, with an object fading from view in less than a second. Indestructible observers plunging into a black hole, on the other hand, are unaffected by any of these consequences once they cross the event horizon. They cross the event horizon after a finite period, according to their own clocks, which appear to tick properly; in classical general relativity, due to Einstein's equivalence principle, it is impossible to establish the position of the event horizon from local observations. At equilibrium, the topology of a black hole's event horizon is always spherical. The event horizon of non-spinning (static) black holes is perfectly spherical, whereas the event horizon of revolving black holes is oblate.

Singularity: A gravitational singularity, a region where the spacetime curvature becomes infinite, may exist in the heart of a black hole, as described by general relativity. This region takes the shape of a single point in a non-spinning black hole, and it is smeared out to create a ring singularity in the plane of rotation in a revolving black hole.

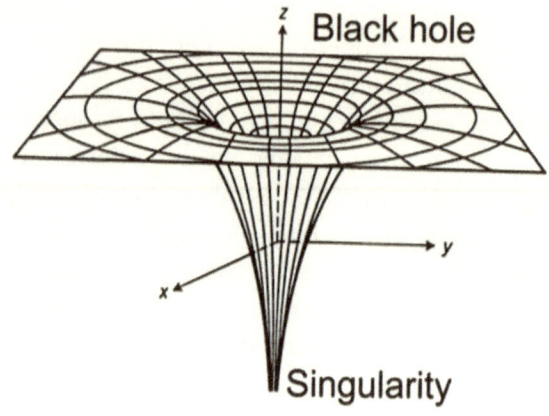

Gravitational singularity, a region where the spacetime curvature becomes infinite

The singular region has zero volume in both circumstances. It's also possible to establish that the solitary region includes all of the black hole solution's mass. As a result, the singular region might be considered to have an infinite density. Observers who fall into a Schwarzschild black hole, which is non-rotating and uncharged, will be dragged into the singularity as they cross the event horizon. They can extend the experience by speeding up to slow their descent, but only to a certain point.

They are crushed to infinite density as they approach the singularity, and their mass is added to the black hole's entire mass. They will have been torn apart by the enormous tidal forces before that happens, a process known as spaghettification or the "noodle effect." It is possible to avoid the singularity in the event of a charged or revolving black hole. Extending these answers as far as they can go reveals the hypothetical possibility of escape the black hole through a wormhole into a new spacetime. Traveling to another universe, on the other hand, is merely a theoretical possibility because any disruption would negate it.

Photon sphere:

The photon sphere is a zero-thickness spherical boundary in which photons traveling along tangents to the sphere are imprisoned in a circular orbit around the black hole. The photon sphere has a radius of 1.5 times the Schwarzschild radius for non-rotating black holes. Their orbits would be dynamically unstable, so any minor disruption, such as a particle of falling matter, would generate an instability that would build over time, either causing the photon to escape the black hole or spiraling inward, eventually crossing the event horizon. While light can still escape from the photon sphere, any light traveling in the other direction will be trapped by the black hole. As a result, any light emitted by items between the photon sphere and the event horizon that reaches an outside observer must have been emitted by objects between the photon sphere and the event horizon.

Ergosphere:

The ergosphere, an area of spacetime in which it is impossible to stand still, surrounds rotating black holes. General relativity predicts that any rotating mass will tend to "drag" along the spacetime immediately surrounding it; this is the outcome of a process known as frame-dragging. Any object that comes close to a rotating mass will begin to rotate in the same direction as the revolving mass. This effect is so powerful at the event horizon for a rotating black hole that an object would have to move faster than the speed of light in the opposite direction to simply remain still. The ergosphere of a black hole is a volume defined by the event horizon of the black hole and the ergosurface, which corresponds with the event horizon at the poles but is significantly farther away around the equator.

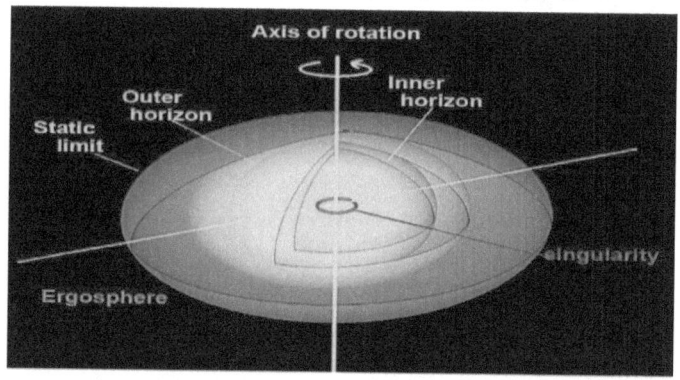

Hawking predicted in 1974 that black holes aren't completely black and that they release modest amounts of thermal radiation. Hawking radiation is the name given to this phenomenon. He discovered that a black hole should emit particles with a perfect black body spectrum by applying quantum field theory to a static black hole background. Since Hawking's publication, numerous people have independently verified the result using a variety of methods. If Hawking's theory of black hole radiation is right, black holes should shrink and evaporate over time as their mass is lost by photon and another particle emission.

The LIGO gravitational wave observatory produced the first successful direct observation of gravitational waves on September 14, 2015. The gravitational waves created by the merger of two black holes, one with 36 solar masses and the other with 29 sun masses, produced a signal that was consistent with theoretical predictions. This discovery is the most conclusive proof of the existence of black holes to date.

On April 10, 2019, an image of a black hole was released, which was enlarged because of the highly twisted light paths near the event horizon. The dark shadow in the middle is caused by the black hole absorbing light pathways. The image is in false color because the observed light halo in this image is radio waves, not visible light.

The Event Horizon Telescope (EHT) is a live program that monitors the immediate environs of black hole event horizons, such as the one at the center of the Milky Way. EHT began observing the black hole at the heart of Messier 87 in April 2017. "In total, eight radio observatories on six mountains and four continents studied the galaxy in Virgo on and off for ten days in April 2017" in order to provide the data that led to the image two years later in April 2019. EHT has released the first direct photograph of a black hole, namely the supermassive black hole in the center of the aforementioned galaxy, after two years of data processing.

WORM HOLE

CHAPTER TEN

WORMHOLE

A wormhole, also known as an Einstein–Rosen bridge or an Einstein–Rosen wormhole, is a hypothetical structure based on a specific solution of the Einstein field equations that connect distant places in space-time.

In connection with mass analysis of electromagnetic field energy, German mathematician, philosopher, and theoretical physicist Hermann Weyl proposed a wormhole hypothesis of matter in 1928; however, he did not use the term "wormhole" – instead, he spoke of "one-dimensional tubes." Instead, American theoretical physicist John Archibald Wheeler, inspired by Weyl's work, coined the term "wormhole" in a 1957 paper co-authored with.

A wormhole might theoretically connect extremely great distances like a billion light-years, tiny distances like a few meters, different points in time, or even separate universes. Wormholes are consistent with Einstein's general theory of relativity, but whether they exist is unknown.A wormhole can be imagined as a tunnel with two endpoints at distinct places in spacetime, such as different locations, times, or both. Space can be viewed as a two-dimensional surface to simplify the concept of a wormhole. In this example, a wormhole would appear as a hole in that surface, lead into a 3D tube on the interior surface of a cylinder, and then re-emerge with a hole comparable to the entry on the 2D surface. A true wormhole would be similar to this, but with one additional spatial dimension.

Another technique to see wormholes is to take a sheet of paper and draw two somewhat distant dots on one side. The sheet of paper represents a plane in the space-time continuum, and the two points represent a distance to be traversed, but a wormhole might hypothetically be created by folding that plane (i.e. the paper) so that the points are touching. Because the two points are now touching, traversing the gap will be considerably easier.

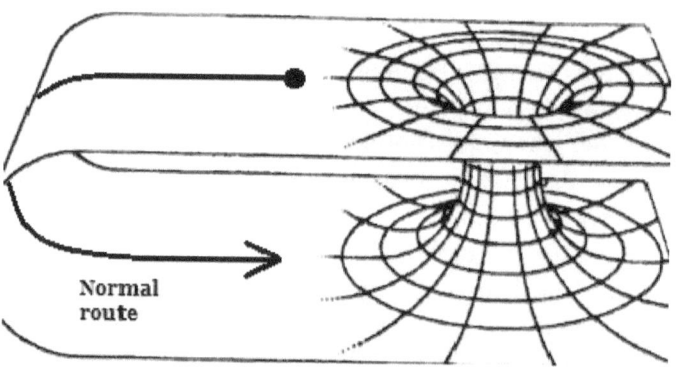

Normal
route

Depiction of a Wormhole

An intra-universe wormhole connecting two places in the same universe is a compact region of spacetime with a topologically trivial boundary but an interior that is not simply connected.Geometrically, wormholes can be described as regions of spacetime that constrain the incremental deformation of closed surfaces.

Types of Wormhole:

The Schwarzschild wormhole, which would be present in the Schwarzschild metric describing an eternal black hole, was the first form of wormhole solution identified, however, it was discovered that it would collapse too quickly for anything to cross from one end to the other. Traversable wormholes, or wormholes that can be crossed in both directions, were thought to be viable only if exotic materials with a negative energy density could be used to stabilize them.

Physicists later stated that tiny traversable wormholes may be conceivable without the use of exotic matter, requiring just electrically charged fermionic matter with a mass small enough to avoid collapsing into a charged black hole.Lorentzian traversable wormholes would allow rapid travel in both directions from one section of the universe to another inside the same universe, as well as transit across universes. In general relativity, the idea of traversable wormholes was first established in a 1973 study by Homer Ellis and independently by K. A. Bronnikov in a 1973 publication.

If traversable wormholes exist, they might allow time travel. A suggested time-travel machine based on a traversable wormhole may theoretically work like this: The wormhole's one end is accelerated to a substantial fraction of the speed of light, maybe via an advanced propulsion technology, and then returned to its birthplace. Another option is to move one of the wormhole's entrances into the gravitational field of an object with higher gravity than the other entrance, then return it to a place near the other opening.

Construction of a traversable wormhole, according to current hypotheses on the physics of wormholes, would necessitate the creation of a substance with negative energy, sometimes referred to as "exotic matter." More precisely, the wormhole spacetime necessitates an energy distribution that defies several energy requirements, including the null energy condition, as well as the weak, strong, and dominant energy conditions.

The impossibility of relative speed faster than light only applies locally. Wormholes could enable superluminal (faster-than-light) travel by ensuring that the speed of light is never exceeded locally. Subluminal (slower-than-light) speeds are used when going via a wormhole.

If a wormhole connects two places with a length shorter than the distance between them outside the wormhole, the time it takes to cross it could be less than the time it would take a light beam to travel through space outside the wormhole. A light beam traveling through the same wormhole, on the other hand, would outrun the traveler.

While it appears that nature does not permit the existence of macroscopic wormholes at this moment, there is still enough uncertainty in the arguments to allow theoretical physicists to continue researching this strange and fascinating element of space-time.

Even if wormholes could be discovered, today's technology is insufficient to enlarge or stabilize them. However, scientists are continuing to research the concept as a means of space travel in the hopes that technology will be able to use it in the future.

DARK MATTER

CHAPTER ELEVEN

DARK MATTER

Dark matter is called dark because it does not appear to interact with the electromagnetic field, which means it does not absorb, reflect or emit electromagnetic radiation, and is therefore difficult to detect.Dark matter is a hypothesized kind of matter that is thought to account for approximately 23-27% of the total mass-energyof the universe.

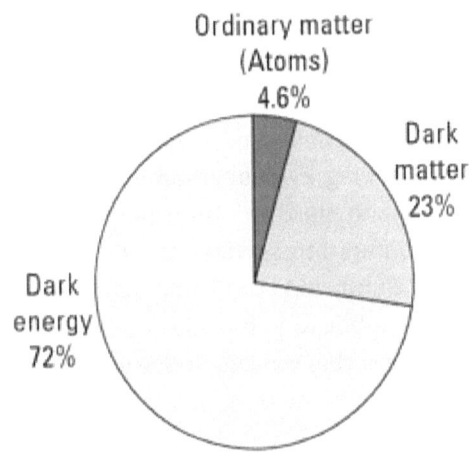

Percentage of Dark matter, Dark energy, and ordinary matter

Observations of gravitational lensing and the cosmic microwave background, as well as astronomical observations of the observable universe's current structure, galaxy formation and evolution, mass location during galactic collisions, and galaxies' motion within galaxy clusters, are all taken as evidence for Dark matter.As the dark matter has not yet been observed directly, if it exists, it must barely interact with ordinary baryonic matter and radiation, except through gravity.

Dark matter has a long and illustrious history. Lord Kelvin calculated the number of dark bodies in the Milky Way from the measured velocity dispersion of the stars revolving around the galaxy's core in a discussion given in 1884. He calculated the mass of the galaxy using these data, which he determined to be different from the mass of visible stars. "Many of our stars, maybe a huge proportion of them, maybe black bodies," Lord Kelvin concluded. Henri Poincaré coined the term "dark matter" in his 1906 book "The Milky Way and Theory of Gases."

Vera Rubin, Kent Ford, and Ken Freeman's work in the 1960s and 1970s, which also used galaxy rotation curves, gave more convincing evidence. Rubin and Ford used a new spectrograph to more precisely quantify the velocity curve of edge-on spiral galaxies. In 1978, this outcome was confirmed. Rubin and Ford's findings were published in an influential publication in 1980.

They discovered that most galaxies must have around six times the amount of dark matter as visible mass; hence, by 1980, the apparent necessity for dark matter had been generally acknowledged as a major unsolved astronomical problem. In the 1980s, a lot of data, including gravitational lensing of background objects by galaxy clusters, the temperature distribution of hot gas in galaxies and clusters, and the pattern of anisotropies in the cosmic microwave background, all supported the existence of dark matter.

Dark matter, according to cosmologists, is mostly made up of a form of subatomic particle that has yet to be identified. One of the key initiatives in particle physics is the search for this particle using various methods.

Dark matter is any substance that interacts with visible matter primarily through gravity, such as stars and planets. As a result, it doesn't have to be built up entirely of new fundamental particles; it might be made up of normal baryonic matter, such as protons or neutrons, at least in part. Most scientists, however, believe that dark matter is dominated by a non-baryonic component, which is likely formed of a currently unknown fundamental particle or comparable exotic state.

Axions, sterile neutrinos, weakly interacting massive particles (WIMPs), gravitationally-interacting massive particles (GIMPs), supersymmetric particles, geons, and primordial black holes are all candidates for non-baryonic dark matter.

Because their individual masses – however uncertain they may be – are nearly definitely too small, the three neutrino types previously discovered can only supply a small proportion of dark matter, given to constraints determined from large-scale structure and high-redshift galaxies.

There are three types of dark matter: cold, warm, and hot. These categories pertain to velocity rather than temperature, indicating how far corresponding items traveled due to random motions in the early universe before slowing due to cosmic expansion - this is the free streaming length, which is a significant distance (FSL).

Cold dark matter causes the structure to form from the bottom up, with galaxies forming first and galaxy clusters afterward, whereas hot dark matter causes the structure to form from the top-down, with huge matter aggregations forming first and later fragmenting into separate galaxies.

Warm dark matter is made up of particles with an FSL similar to that of a protogalaxy. Warm dark matter predictions are similar to cold dark matter predictions on large scales, but with fewer small-scale density disturbances.This reduces the predicted abundance of dwarf galaxies and may lead to a lower density of dark matter in the central parts of large galaxies.

Warm dark matter is a category that no known particles fit into. The sterile neutrino is one candidate: a heavier, slower form of neutrino that, unlike other neutrinos, does not interact through the weak interaction. To make their equations work, some modified gravity theories, like scalar-tensor–vector gravity, require "warm" dark matter.

Most cosmic observations can be explained simply by cold dark matter. It's dark matter made up of elements whose FSL is substantially lower than that of a protogalaxy.

Because hot dark matter does not appear to be capable of enabling galaxy or galaxy cluster formation, and most particle candidates slowed early, this is the focus of dark matter study.

Evidence from observation:

Structure formation:

The era after the Big Bang when density perturbations condensed to form stars, galaxies, and clusters is referred to as structure creation. The Friedmann solutions to general relativity describe a homogeneous cosmos before structure development. Small anisotropies grew and condensed the homogenous universe throughout time, eventually forming stars, galaxies, and larger structures. Radiation, which was the primary element of the cosmos at the beginning, affects ordinary matterAs a result, its density disturbances are wiped out, making it impossible for them to condense into the structure. There would not have been enough time for density perturbations to form into the galaxies and clusters visible today if there was an only ordinary matter in the cosmos.Because it is unaffected by radiation, dark matter offers a solution to this problem. As a result, its density perturbations will be the first to grow. The gravitational potential that results in functions as an attractive potential well for ordinary matter to collapse later, speeding up the structure creation process.

Curves of galaxy rotation

Spiral galaxies have arms that spin around the galactic core. As one moves from the center to the edges of a spiral galaxy, the luminous mass density diminishes. We can model the galaxy as a point mass in the center with test masses orbiting around it, similar to the Solar System if the luminous mass is all there is. According to Kepler's Second Law, rotation velocities should decrease with distance from the center, as they do in the Solar System. This is not the case. Instead, as the distance from the center grows, the galaxy's rotation curve remains flat. If Kepler's laws are correct, then the obvious way to resolve this discrepancy is to conclude the mass distribution in spiral galaxies is not similar to that of the Solar System. In particular, there is a lot of non-luminous matter -dark matter in the outskirts of the galaxy.

Gravitational lensing:

Massive objects (such as a cluster of galaxies) situated between a more distant source (such as a quasar) and an observer should operate as a lens to bend the light from this source, according to general relativity. The greater the bulk of an object, the more lensing is seen. The observed bending of background galaxies into arcs when their light passes through such a gravitational lens is known as strong lensing. It's been seen in the vicinity of several distant clusters, notably Abell 1689. The mass of the intervening cluster can be determined by measuring the distortion geometry. In the dozens of cases where this has been done, the mass-to-light ratios obtained correspond to the dynamical dark matter measurements of clusters.

Multiple copies of a picture can result via lensing. Scientists were able to derive and map the distribution of dark matter around the MACS J0416.1-2403 galaxy cluster by examining the dispersion of several picture copies. The Dark Energy Survey Collaboration has released a new detailed dark matter map. Using machine learning technology, the atlas also showed previously unknown filamentary structures connecting galaxies.

Dark matter does not bend light itself; mass (in this case the mass of the dark matter) bends space-time. Light follows the curvature of space-time, resulting in the lensing effect.

Cosmic microwave background:

Although dark matter does not interact directly with radiation, it has an impact on the CMB through its gravitational potential and its influence on the density and velocity of ordinary matter on enormous scales. As a result, ordinary and dark matter perturbations evolve at different rates and leave different impressions on the cosmic microwave background (CMB).

The cosmic microwave background is quite close to a perfect blackbody, but there are a few parts in 100,000 temperature anisotropies. An angular power spectrum, which is observed to feature a succession of acoustic peaks at almost similar spacing but varying heights, can be deconstructed from a sky map of anisotropies.

Modern computer programs such as CMBFAST and CAMB can forecast the series of peaks for any assumed set of cosmic parameters, and therefore matching theory to data constrains cosmological parameters. The density of baryonic matter is largely shown in the first peak, while the density of dark matter is mostly shown in the third peak, which measures the density of matter and the density of atoms. The observed CMB angular power spectrum provides powerful evidence in support of dark matter.

Dark matter particle detection:

If dark matter is made up of subatomic particles, millions, if not billions, must pass through every square centimeter of the Earth every second. Although WIMPs are popular search candidates, the Axion Dark Matter Experiment (ADMX) searches for axions. Another candidate is heavy hidden sector particles, which only interact with ordinary matter via gravity. These experiments can be divided into two classes: direct detection experiments, which search for the scattering of dark matter particles off atomic nuclei within a detector; and indirect detection, which look for the products of dark matter particle annihilations or decays.

Because dark matter has yet to be definitively found, a variety of other theories have evolved to explain the phenomena that dark matter was supposed to explain. Modifying general relativity is the most usual way.

On solar system scales, general relativity has been thoroughly verified, but its validity on galactic or cosmological scales has yet to be established. A proper change to general relativity may eliminate the necessity for dark matter. MOND and its relativistic generalization tensor-vector-scalar gravity (TeVeS), f(R) gravity, negative mass, dark fluid, and entropic gravity are the most well-known theories in this category.

As important as dark matter is thought to be in the cosmos, direct evidence of its existence and a thorough knowledge of its nature have eluded scientists. Alternative theories are proposed to explain the anomalies in observed galaxy rotation, while dark matter remains the most frequently accepted theory. None of these ideas, on the other hand, has received universal acceptance among scientists.

Alternative ideas have the drawback of observable evidence for dark matter coming from so many different sources. It is possible to explain any individual observation, but it is extremely difficult to explain all of them in the absence of dark matter.

Most astrophysicists believe that while changes to general relativity could theoretically explain some of the observable evidence, there is likely enough evidence to conclude that there must be some type of dark matter existent in the Universe.

Researchers have constructed the world's largest map of dark matter. Astronomers can track the existence of matter by looking at light traveling to Earth from faraway galaxies because stuff curves space-time. If the light has been warped, there is something in the foreground that is bending the light as it approaches us.

Artificial intelligence was utilized by the Dark Energy Survey (DES) team to examine photos of 100 million galaxies to see if they had been stretched. The new map, seen above, depicts all matter discovered in the foreground of known galaxies and spans a quarter of the southern hemisphere sky.

DARK ENERGY

CHAPTER TWELVE

DARK ENERGY

Dark energy is an undiscovered kind of energy that impacts the cosmos on the biggest sizes, according to physical cosmology and astronomy. Measurements of supernovae provided the first observational evidence for its existence, demonstrating that the cosmos does not expand at a steady rate, but rather accelerates.

Dark energy's nature is more hypothetical than dark matter's, and many aspects of it remain a subject of conjecture. Dark energy is assumed to be relatively homogeneous and not particularly dense, and it is not known to interact with any of the fundamental forces save gravity. It is unlikely to be detected in laboratory investigations since it is highly rarefied and non-massive. Because it consistently fills otherwise space, dark energy may have such a tremendous effect on the universe, accounting for 68 percent of universal density despite its dilute nature.

All types of matter and energy in the universe were considered to cause the expansion of the universe to slow down over time. The cosmic microwave background suggests that the cosmos originated with a hot Big Bang, from which general relativity explains the universe's history and subsequent large-scale motion. There was no way to explain how an accelerating universe could be measured without introducing a new kind of energy. Dark energy has been the most widely accepted explanation for the increased expansion since the 1990s.

Dark matter and conventional baryonic matter provide 26% and 5% of the mass energy, respectively, whereas other components like neutrinos and photons contribute very little. Within galaxies, the density of dark energy is substantially lower than the density of ordinary matter or dark matter. However, because it is homogeneous across space, it dominates the universe's mass-energy balance.

Evidence of existence

The evidence for dark energy is circumstantial, but it comes from three different places: The relationship between distance measurements and redshift, which suggests the cosmos has expanded more in the second part of its age.

The absence of any discernible global curvature necessitates the theoretical necessity for a type of extra energy that is not matter or dark matter to construct the observationally flat universe. Measures of large-scale wave patterns of mass density in the universe. Recent observations of supernovae point to a cosmos that is 71.3 percent dark energy and 27.4% dark matter and baryonic matter.

To reconcile the measured geometry of space with the total amount of matter in the universe, the existence of dark energy, in whatever form it takes, is required. Anisotropies of the cosmic microwave background (CMB) imply that the universe is nearly flat, according to measurements. The mass-energy density of the cosmos must be equal to the critical density for the universe to be flat. According to the CMB spectrum, the total amount of matter in the universe (including baryons and dark matter) accounts for only roughly 30% of the critical density. This implies that there is another source of energy that accounts for the remaining 70%.

According to the theory of large-scale structure, which governs the creation of structures in the universe (stars, quasars, galaxies, galaxy groups, and clusters), the density of matter in the universe is only 30% of the critical density.

The evidence for dark energy is primarily reliant on general relativity theory. As a result, a change to general relativity may also eliminate the necessity for dark energy. There are numerous theories of this nature, and study is ongoing. Many modified gravity ideas as explanations for dark energy were ruled out by the measurement of the speed of gravity in the first gravitational wave observed by non-gravitational techniques.

According to cosmologists, acceleration began some 5 billion years ago. Previously, it was considered that the growth was slowing down due to the alluring attraction of matter. Dark matter's density diminishes faster than dark energy's in an expanding universe, and dark energy finally takes over. The density of dark matter is halved when the volume of the universe doubles, whereas the density of dark energy remains essentially unaltered, it is exactly constant in the case of a cosmological constant.

Other, more speculative beliefs regarding the universe's future exist based on Dark energy. The phantom energy model of dark energy leads to divergent expansion, implying that dark energy's effective force grows until it outnumbers all other forces in the universe. Dark energy will gradually take away all gravitationally bound objects, including galaxies and solar systems, and eventually overcome the electrical and nuclear forces to break apart atoms, terminating the universe in a "Big Rip," according to this scenario.

Models of Dark energy

The density of dark energy may have changed over time in the universe's history. We can estimate the current density of dark energy using modern observational data. It is feasible to explore the effect of dark energy on the history of the Universe and restrict parameters of the dark energy equation of state using baryon acoustic oscillations. Several models have been presented to this end. The Chevallier–Polarski–Linder model is one of the most popular (CPL). Barboza & Alcaniz. 2008, Jassal et al. 2005,Wetterich. 2004 (Oztas et al. 2018), are some more common models.

Inhomogeneous cosmology, for example, is an alternative to dark energy that aims to explain observational data. Dark energy, in this situation, does not exist and is only a measuring artifact. For example, if we are in an area of space that is less dense than typical, the measured cosmic expansion rate could be misinterpreted as a change in time or acceleration. A different technique employs a cosmic application of the equivalence principle to demonstrate how space appears to be expanding more rapidly in the gaps surrounding our local cluster. While small, such effects could add up over billions of years to become significant.Another possibility is that the universe's fast expansion is a mirage generated by our relative speed to the rest of the cosmos, or that the statistical methods utilized were faulty.

Dark Energy alternatives have been offered. According to some scientists, our Galaxy is located within a low-density area induced by the passage of a density wave. This large-scale wave in space-time could have been caused by the Big Bang. As this primordial wave travelled through the universe, it left a low-density ripple tens of millions of light-years across, which is now home to the Galaxy. While this discrepancy in space-time features is theoretically possible, it would violate the Copernican principle, which claims that the universe is homogeneous on enormous scales.

Dark energy – also known as the cosmological constant or quintessence – is the most widely accepted theory, backed up by plenty of empirical data. NASA's orbiting WFIRST telescope and the international Dark Energy Survey, both based in Chile, are two present and future space projects and ground-based surveys that will examine the nature of dark energy.

FATE OF THE UNIVERSE

CHAPTER THIRTEEN

FATE OF THE UNIVERSE

Choosing the fate and evolution of the universe based on current observational facts has become a genuine cosmological question. The average movements of galaxies, the form and structure of the universe, and the amount of dark matter and dark energy in the universe are all factors to consider when calculating the universe's origin and eventual fate.

The Big Crunch

The Big Crunch is a hypothetical scenario for the universe's ultimate fate, in which the universe's expansion finally reverses and re-collapses, causing the cosmic scale factor to approach zero, potentially followed by a reformation of the cosmos beginning with another Big Bang. The overwhelming majority of evidence contradicts this viewpoint. Rather than being slowed by gravity, astronomical studies show that the expansion of the cosmos is speeding, implying that the universe is significantly more likely to terminate in heat death or a Big Rip.

Physicist Roger Penrose proposed the conformal cyclic cosmology, a general relativity-based theory in which the cosmos expands until all matter decays and is converted to light. Nothing in the universe would have a time or space scale linked with it, therefore the Big Bang becomes synonymous with it. A more specific hypothesis known as "Big Bounce" argues that the cosmos might collapse to its original condition and then begin a new Big Bang, allowing the universe to endure indefinitely while passing through stages of expansion. Contraction and the Big Bang are a large crunch.

Paul Davies thought of a scenario in which the Big Crunch occurs 100 billion years from now. In his model, the contracting world would progress in the same way that the expanding universe did. Galaxy clusters, then galaxies, would merge first, raising the temperature of the cosmic microwave background (CMB) as CMB photons became blue shifted.

Stars would eventually become so close to one another that they would collide. When the CMB grows hotter than M-type stars, they lose their ability to radiate heat and cook themselves until they evaporate; this process continues for successively hotter stars until O-type stars boil away roughly 100,000 years before the Big Crunch. The temperature of the cosmos would be so high in the final minutes that atoms and atomic nuclei would break apart and be dragged into already coalescing black holes.

All matter in the cosmos would be crushed into an endlessly hot, indefinitely dense singularity identical to the Big Bang at the time of the Big Crunch. According to the Big Crunch scenario, the density of matter throughout the universe is sufficiently enough that gravitational attraction will overpower the Big Bang's expansion.

The overwhelming majority of evidence contradicts this viewpoint. Rather than being slowed by gravity, astronomical studies show that the expansion of the cosmos is speeding, implying that the universe is significantly more likely to terminate in heat death or a Big Rip.

The Big Rip

The Big Rip is a hypothetical cosmological model concerning the universe's ultimate fate, in which the universe's matter, from stars and galaxies to atoms and subatomic particles, and even spacetime itself, is progressively torn apart by the universe's expansion at some point in the future until distances between particles become infinite. The scale factor of the universe is increasing, according to the standard model of cosmology, and will increase exponentially in the future epoch of cosmological constant dominance.

The hypothesis's validity is contingent on the type of dark energy present in our universe. A type of dark energy known as phantom energy, which is always expanding, could confirm this notion. If the amount of dark energy in the universe continues to grow inexorably, it will be able to overcome all of the forces that keep the universe together.

A cosmos dominated by phantom energy expands at a faster rate than the rest of the universe. However, this indicates that the observable universe and particle horizon are shrinking all the time - the distance at which objects traveling at the speed of light from an observer

" or "Big

)e of dark
er hand, is
re of dark
·k energy,
:ct on the
; plausible
ary era in
pacts are
properly

ırk matter
;s for their
unknown.
·e certain.

e over which
r and shorter.
he size of any
the structure's
to any of the
orn apart."The
model predicts
l singularity is
in which the
es.
ie day witness
on them. The
ripped apart,
wn to quarks

f the universe,
ant, a constant
usly, then the
ng rate.
orably darken
cease to light.
ar remains left
ig only black
emit Hawking

will continue
ine prominent
is it expands,
survive. As a
own as "Heat

THEORY OF EVERYTHING

CHAPTER FOURTEEN

THEORY OF EVERYTHING

A theoretical framework of physics known as a theory of everything TOE, a final and ultimate theory, is a hypothetical, singular, all-encompassing, theoretical framework that fully explains and links together all physical aspects of the universe.

As theories of everything, String theory and M-theory have been presented. Two theoretical frameworks have emerged over the centuries that, when combined, most closely resemble a TOE. General relativity and quantum mechanics are the two ideas that underpin all current physics. General relativity is a theoretical framework that focuses solely on gravity to describe the cosmos in large-scale and high-mass regions such as stars, galaxies, and clusters of galaxies.

Quantum mechanics, on the other hand, is a theoretical framework for describing the world in small-scale and low mass regions that only concentrates on three non-gravitational forces: subatomic particles, atoms, molecules, and so on. Quantum mechanics has effectively implemented the Standard Model, which describes all observable basic particles as well as the three non-gravitational forces — strong nuclear, weak nuclear, and electromagnetic force.

In their respective fields of application, general relativity and quantum mechanics have been thoroughly proven. Because the common applicability regions of general relativity and quantum mechanics are so dissimilar, most situations necessitate the use of only one of the two theories. In regions of extremely small scale that are in the Planck scale – such as those found within a black hole or during the early phases of the universe, the two theories are thought to be irreconcilable.

To resolve the incompatibility, a theoretical framework must be discovered that reveals a deeper underlying reality, unifying gravity with the other three interactions, and harmoniously integrates the realms of general relativity and quantum mechanics into a seamless whole: the TOE is a single theory capable of describing all phenomena in the universe.All of nature's fundamental interactions would be unified by a Theory of Everything, which would include gravitation, the strong interaction, the weak interaction, and electromagnetic. Because the weak interaction can change the nature of elementary particles, the TOE should be able to anticipate all of the many types of particles that are feasible.

To unify electromagnetism and the weak and strong forces, several Grand Unified Theories (GUTs) have been proposed. Grand unification would require the existence of an electronuclear force, which is expected to manifest at energies of the order of 1016 GeV, much beyond the current capabilities of any particle accelerator. Although the simplest GUTs have been ruled out empirically, the concept of a grand unified theory, especially when coupled with supersymmetry, remains a popular candidate among theoretical physicists.

Quantum gravity has become one field of ongoing research in the pursuit of this objective. String theory, for example, has become a candidate for the TOE.String theory research has been aided by several theoretical and experimental reasons.It has started to address some of the key concerns in quantum gravity, such as addressing the black hole information paradox, counting the correct entropy of black holes, and allowing for topology-changing events on the theoretical side. The theory of causal fermion systems, which uses the two present physical theories general relativity and quantum field theory as limiting instances, is one of several attempts to construct a theory of everything. Causal Sets is a different theory. Its direct goal, like some of the other techniques discussed above, isn't necessarily to create a TOE, but rather to develop a workable quantum gravity theory that may someday contain the standard model and become a contender for a TOE.

Another attempt could be linked to ER=EPR, a physicist's claim that entangled particles are linked through a wormhole or Einstein–Rosen bridge and it could serve as a foundation for combining general relativity and quantum physics into a universal theory.

As of now, there is no candidate theory of everything that incorporates the standard model of particle physics and general relativity while also being able to calculate the fine-structure constant or electron mass. Most particle physicists think that the outcome of ongoing experiments – the search for new particles at the giant particle accelerators and dark matter – is needed to offer further input for a TOE.

Aside from the pure intellectual joy of finishing a centuries-long pursuit, past examples of unification have anticipated new phenomena, some of which have proven to be of tremendous practical relevance, such as electrical generators. And, as with these previous unification examples, the TOE would most likely allow us to securely specify the domain of validity and residual error of low-energy approximations to the entire theory.

The layers of nature, according to Frank Close - a particle physicist, may be analogous to the layers of an onion, with an endless number of layers. This would indicate the existence of an endless number of physical theories.

According to several academics, Gödel's incompleteness theorem implies that any attempt to design a TOE is doomed to fail. Informally stated, Gödel's theorem states that any formal theory strong enough to prove elementary arithmetical facts is either inconsistent (both a statement and its denial can be derived from its axioms) or incomplete, in the sense that there is a true statement that cannot be derived in the formal theory. Because most physicists believe that stating the underlying rules suffices as a definition of a "theory of everything," they claim that Gödel's Theorem does not rule out the possibility of a TOE.

At present, neither unification of the strong and electroweak forces – nor a unification of these forces with gravitation – has been accomplished in the quest for a theory of everything.

To date, no physical hypothesis is thought to be perfectly accurate. Instead, physics has progressed through a succession of "successive approximations," which have allowed for increasingly accurate predictions across a larger spectrum of phenomena. Finding a theory of everything would be a huge accomplishment, allowing us to finally make sense of all the strange and amazing things that exist in our universe. Cosmological theories have been proposed for as long as humans have attempted to make sense of the universe.

Unlike other fields of science, cosmology is unique in that it can only examine one universe. We can't change one parameter, manipulate another, and end up with a completely altered system to test. We will never know how unique our universe is because we have no other universe to compare it to.

HISTORY OF COSMOLOGICAL THEORIES AND DISCOVERIES TILL NOW

CHAPTER FIFTEEN

HISTORY OF COSMOLOGICAL THEORIES AND DISCOVERIES TILL NOW

Mesopotamian cosmology depicts a flat, circular Earth encircled by a cosmic ocean in the 16th century BCE.

15th—11th century BCE – The Hindu Rigveda has certain cosmological hymns, most notably the Nasadiya Sukta, which explains the origin of the cosmos and is derived from the monistic Hiranyagarbha or "Golden Egg."

For 311.04 trillion years, primal stuff is manifest, and for the same amount of time, it is unmanifest. For 4.32 billion years, the universe is manifest, and for an equal amount of time, it is unmanifest. Countless universes exist at the same time. Desires have always powered these cycles, and they will continue to do so indefinitely.

The Babylonian world map depicts the Earth surrounded by the cosmic ocean and seven islands grouped around it to form a seven-pointed star around the sixth century BCE.

Greek thinkers as early as Anaximander establish the concept of many or perhaps infinite universes in the 6th–4th century BCE. Democritus went on to say that the distance and size of these worlds varied, as did the presence, quantity, and size of their suns and moons, and that they were vulnerable to destructive collisions.

Aristotle presents an Earth-centered universe in which the Earth is fixed and the cosmos (or universe) is finite in size but limitless in time in the 4th century BCE. Others, such as Philolaus and Hicetas, were opposed to geocentrism. Plato appears to have argued that there was a beginning to the universe.

5th or earlier century – To the east, there are "hundreds of thousands of billions, boundlessly, incomparably, incalculably, unspeakably, inconceivably, unimaginably, inexplicable many worlds" and "infinite worlds in the ten directions," according to ancient Buddhist teachings.

Several astronomers, including Aryabhata, advocate a Sun-centered cosmology in the 5th–11th century.

964AD – In his Book of Fixed Stars, Persian astronomer Abd al-Rahman al-Sufi (Azophi) makes the first recorded observations of the Andromeda Galaxy and the Large Magellanic Cloud, the first galaxies other than the Milky Way to be sighted from Earth.

In 1543, inhis De revolutionibus orbium coelestium, Nicolaus Copernicus published his heliocentric universe.

1610 — Johannes Kepler uses the night sky to prove that the universe is finite.

Sir Isaac Newton's principles describe large-scale motion in the universe in 1687.

1905 – Albert Einstein publishes his Special Theory of Relativity, which asserts that space and time are not two distinct continua.

Albert Einstein publishes his General Theory of Relativity in 1915, demonstrating how an energy density warps spacetime.

Edwin Hubble measures the distances between the Andromeda Galaxy (M31), the Triangulum Galaxy (M33), and NGC 6822, three neighboring spiral nebulae (galaxies) in 1923. The distances position them well beyond our Milky Way, implying that fainter galaxies are considerably farther out and that the cosmos is made up of thousands of galaxies.Edwin Hubble establishes the linear redshift-distance relationship, demonstrating the universe's expansion in 1929.

In 1948 Based on the behavior of primordial radiation in an expanding universe, George Gamow predicts the presence of cosmic microwave background radiation, and Hermann Bondi, Thomas Gold, and Fred Hoyle propose steady-state cosmologies based on the perfect cosmological principle

In 1950-Fred Hoyle coined the term "Big Bang," claiming that it was not intended to be derogatory; rather, it was intended to underline the distinction between the Big Bang and the Steady-State model.

In 1965 Bell Labs astronomers Arno Penzias and Robert Wilson discover the 2.7 K microwave background radiation.

In 1967 – Robert Wagner, William Fowler, and Fred Hoyle demonstrate that the hot Big Bang accurately predicts the abundances of deuterium and lithium.

In 1980 – Independently, Alan Guth and Alexei Starobinsky suggest the inflationary Big Bang universe as a possible solution to the horizon and flatness issues.

1982 – James Peebles, J. Richard Bond and George Blumenthal argue that cold dark matter is the dominant force in the cosmos.

In 1990 – Preliminary measurements from NASA's COBE mission reveal that the cosmic microwave background radiation has a blackbody spectrum to incredible precision, ruling out the idea of a steady-state enthusiast's proposed integrated starlight model for the background.

In 2002 – Chile's Cosmic Background Imager (CBI) captured photos of the cosmic microwave background radiation with a resolution.

Planck, a European Space Agency (ESA) space observatory, studied the anisotropies of the cosmic microwave background radiation with enhanced sensitivity and modest angular resolution from 2009 to 2013.

The image of the black hole at the heart of the M87 Galaxy is published by the Event Horizon Telescope Collaboration in 2019. [This is the first time scientists have ever taken a photograph of a black hole, proving their existence and assisting in the verification of Einstein's general theory of relativity. This was accomplished by the use of interferometry with a very long baseline.

In 2020 — Scientists publish a paper claiming that the Universe is no longer expanding at the same pace in all directions, implying that the widely held isotropy hypothesis may be incorrect. While prior research has suggested this, this is the first study to look at galaxy clusters in X-rays, which Norbert Schartel believes is of far larger significance. The study discovered a persistent and strong directional tendency of deviations of the normalization parameter A, or the Hubble constant H0, which had previously been regarded as indicating a "crisis of cosmology" by others.

> "I once thought that if I could ask God one question, I would ask how the universe began, because once I knew that, all the rest is simply equations. But as I got older I became less concerned with how the universe began. Rather, I would want to know why he started the universe. For once I knew that answer, then I would know the purpose of my own life.
>
> The basic laws of the universe are simple, but because our senses are limited, we can't grasp them. There is a pattern in creation.
>
> --Albert Einstein"

Epilogue

In an unfathomably enormous Universe, humanity occupies a relatively little spot. We could cross the Milky Way in 100,000 years if we traveled at the speed of light -671 million miles per hour. Even so, we wouldn't have gotten very far. According to the latest estimates, there are 2 trillion galaxies in the observable Universe, and the region they occupy covers at least 90 billion light-years.

The French philosopher Blaise Pascal wrote in Pensées (1669):

> "'When I consider the short duration of my life, swallowed up in eternity before and after, the little space I fill engulfed in the infinite immensity of spaces whereof I know nothing, and which know nothing of me, I am terrified. The eternal silence of these infinite spaces frightens me.'
>
> From beginning to finish, the cosmic point of view embraces everything in the Universe: all of space and time, from edge to edge and beginning to end. We are, at least physically and chronologically, nothing more than a minuscule blip from that perspective."